The Thinking T
SCIENCE HANDBOOK
& PORTFOLIO

CONTENTS:

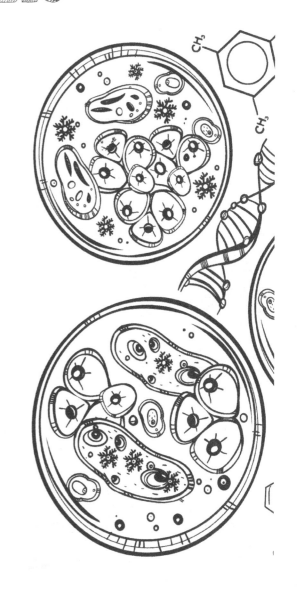

Created By Sarah Janisse Brown & The Thinking Tree LLC

www.FunSchoolingBooks.com

My Science Journal

What do you want to learn about science?

1.

2.

3.

4.

5.

Action Steps:

1. Go to the library or bookstore.

2. Bring home a stack of interesting books about science.

Choose some that have diagrams, instructions and illustrations.

3. Plan to watch science tutorials and do your own research and experiments.

Supplies Needed:

You will need pencils, black drawing pens, science tools, colored pencils, gel pens and markers.

Materials Needed for Projects and Experiments:

_____

_____

_____

_____

_____

_____

_____

Draw the Covers of All Your
SCIENCE BOOKS

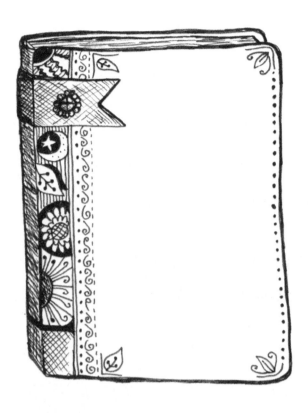
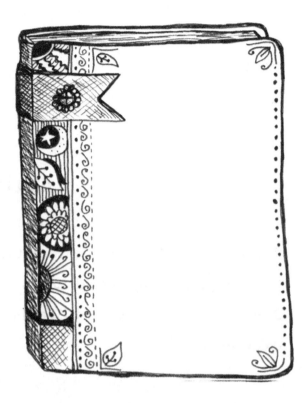
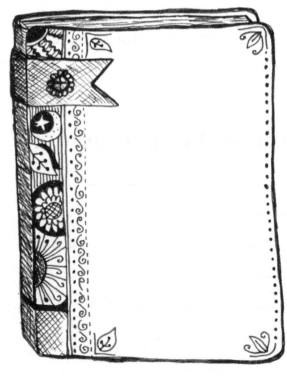

You will use your choice
of science books along with this portfolio.

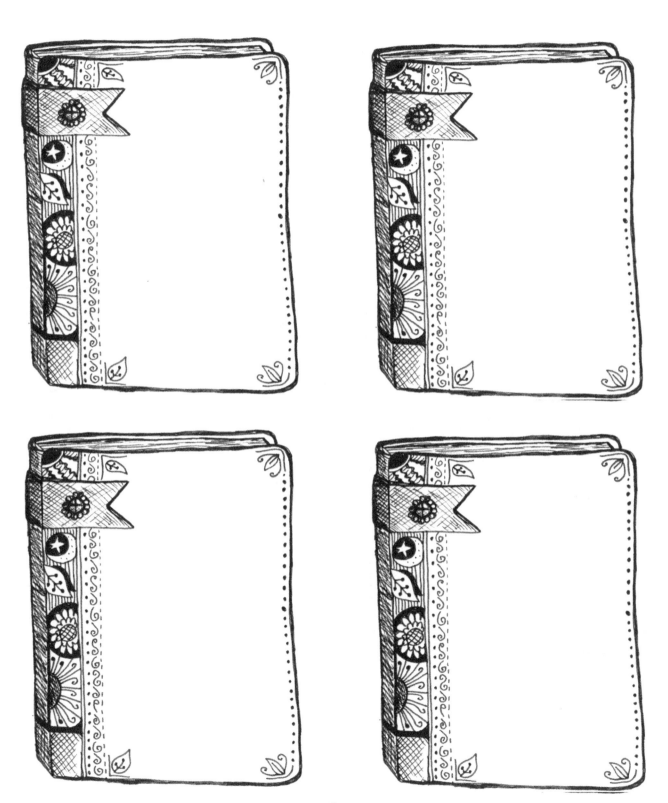

Draw the Covers of All Your
SCIENCE BOOKS

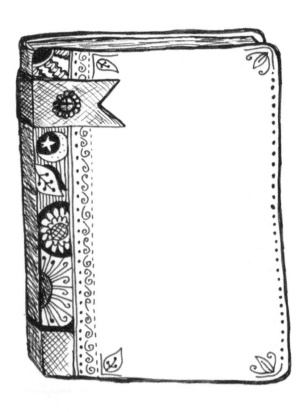

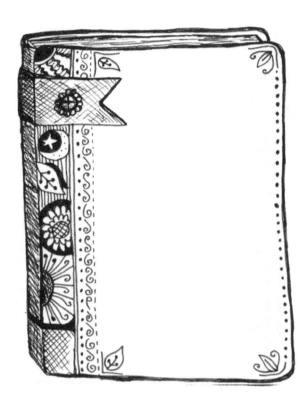

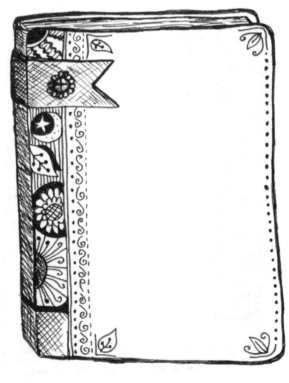

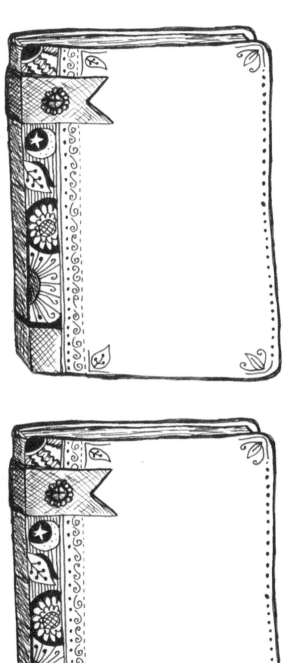

Draw all of your
SCIENCE TOOLS

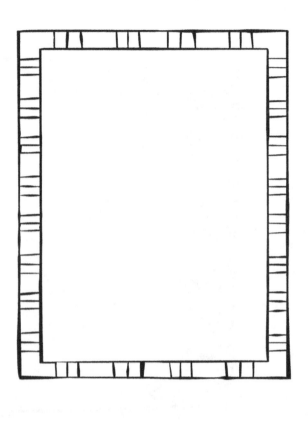

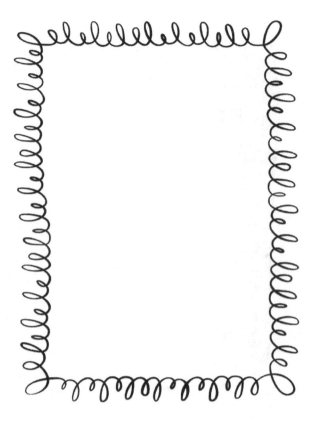

RESEARCH & READING TIME

Name of Book:_____

Topic:_____ Date:_____

Chapter Name: _____ Page Number:____

Title: _____

Science

Drawings & Notes

RESEARCH & READING TIME

Name of Book:_____

Topic:_____ Date:_____

Chapter Name: _____ Page Number:____

Title: _____

Science

Drawings & Notes

RESEARCH & READING TIME

Name of Book:_____

Topic:_____ Date:_____

Chapter Name: _____ Page Number:____

Title: _____

Science

Drawings & Notes

RESEARCH & READING TIME

Name of Book:_____

Topic:_____ Date:_____

Chapter Name: _____ Page Number:____

Title: _____

Science

Drawings & Notes

RESEARCH & READING TIME

Name of Book:_____

Topic:_____ Date:_____

Chapter Name: _____ Page Number:_____

Title: _____

Science

Drawings & Notes

RESEARCH & READING TIME

Name of Book:_____

Topic:_____ Date:_____

Chapter Name: _____ Page Number:_____

Title: _____

Science

Drawings & Notes

RESEARCH & READING TIME

Name of Book:_____

Topic:_____ Date:_____

Chapter Name: _____ Page Number:____

Title: _____

Science

Drawings & Notes

RESEARCH & READING TIME

Name of Book:_____

Topic:_____ Date:_____

Chapter Name: _____ Page Number:____

Title: _____

Science

Drawings & Notes

RESEARCH & READING TIME

Name of Book:_____

Topic:_____ Date:_____

Chapter Name: _____ Page Number:____

Title: _____

Science

Drawings & Notes

MY EXPERIMENTS

(Tape or Draw Images Here)

MY OBSERVATIONS & DISCOVERIES

MY EXPERIMENTS

(Tape or Draw Images Here)

MY OBSERVATIONS & DISCOVERIES

MY EXPERIMENTS

(Tape or Draw Images Here)

MY OBSERVATIONS & DISCOVERIES

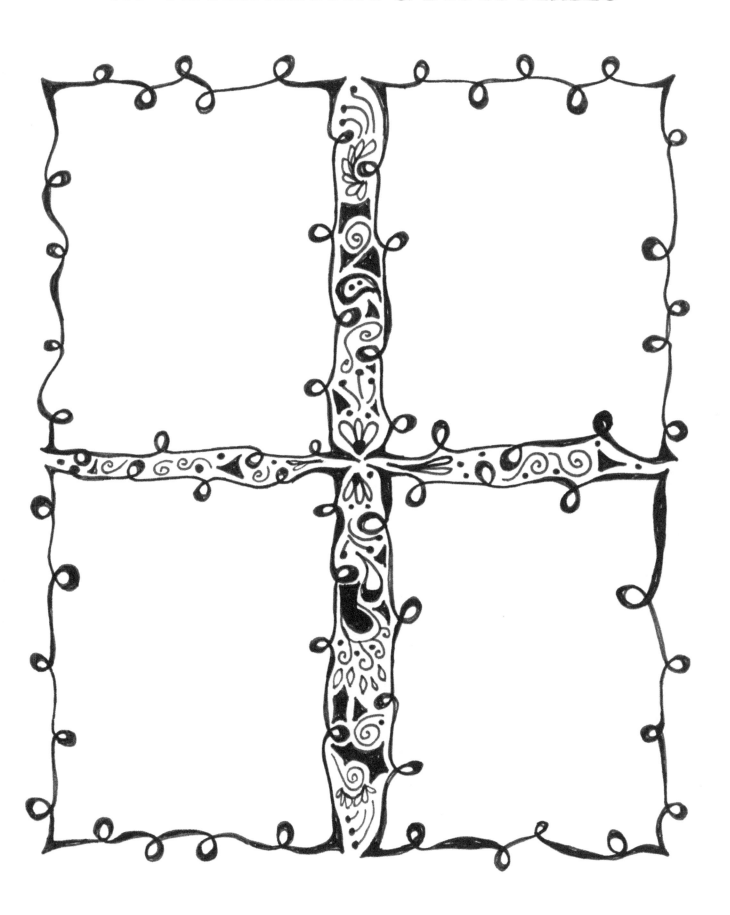

MY EXPERIMENTS

(Tape or Draw Images Here)

MY OBSERVATIONS & DISCOVERIES

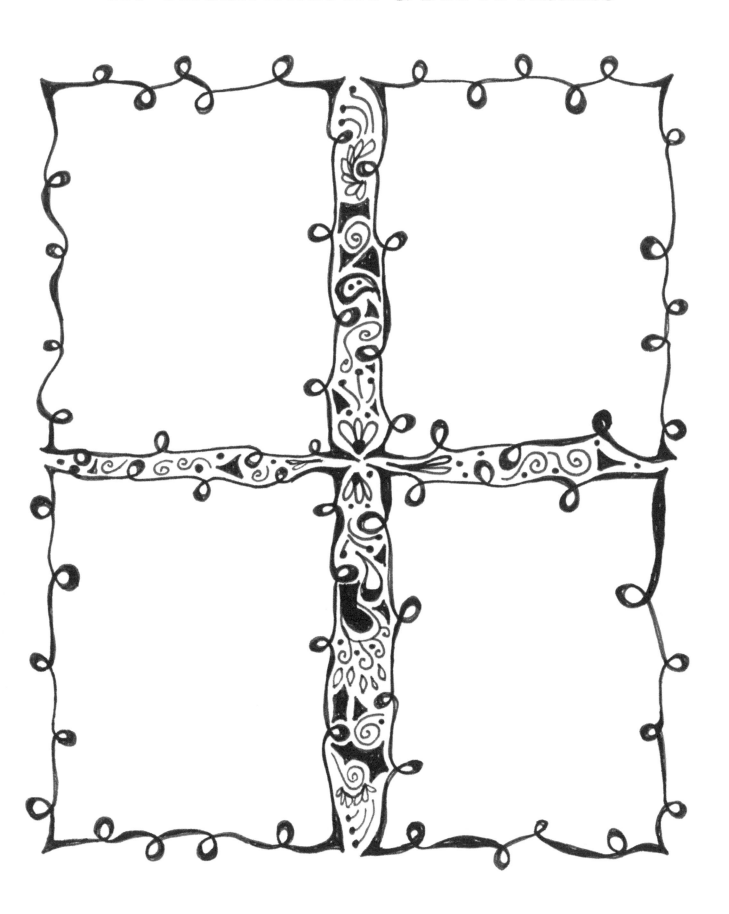

MY EXPERIMENTS

(Tape or Draw Images Here)

MY OBSERVATIONS & DISCOVERIES

BIOGRAPHY OF A SCIENTIST

TITLE:_____

Date_____

BIOGRAPHY OF A SCIENTIST

TITLE:_____

Date_____

BIOGRAPHY OF A SCIENTIST

TITLE:_____

Date_____

BIOGRAPHY OF A SCIENTIST

TITLE:_____

Date_____

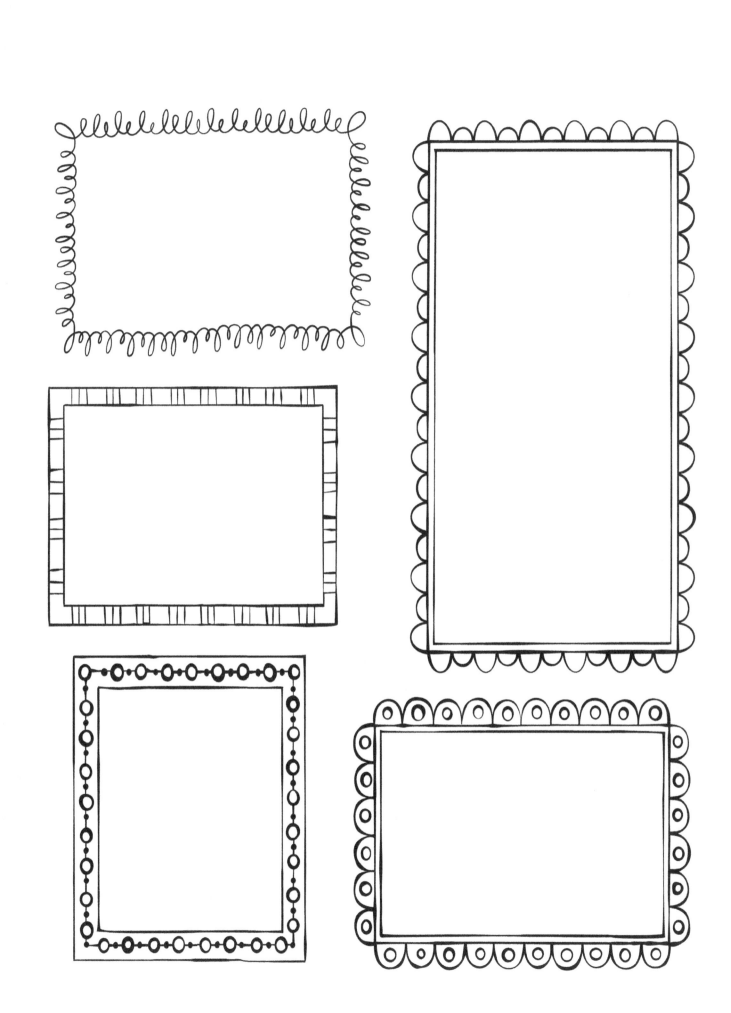

BIOGRAPHY OF A SCIENTIST

TITLE:_____

Date_____

SCIENCE STORIES

If I had Discovered:

_____

A Creative Writing Activity

Drawings, Photos & Images:

SCIENCE STORIES

If I had Discovered:

_____

A Creative Writing Activity

Drawings, Photos & Images:

SCIENCE STORIES

If I had Discovered:

_____

A Creative Writing Activity

Drawings, Photos & Images:

SCIENCE STORIES

If I had Discovered:

A Creative Writing Activity

Drawings, Photos & Images:

SCIENCE STORIES

If I had Discovered:

A Creative Writing Activity

Drawings, Photos & Images:

A STUDY OF SCIENTISTS & SCIENTIFIC DISCOVERIES

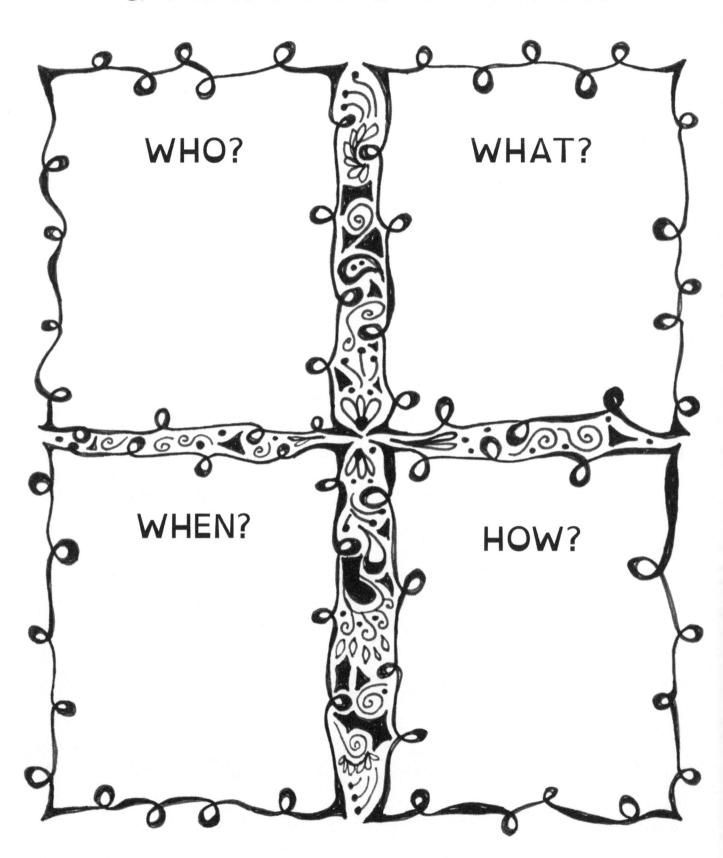

WHO?

WHAT?

WHEN?

HOW?

Drawings, Photos & Images:

A STUDY OF SCIENTISTS & SCIENTIFIC DISCOVERIES

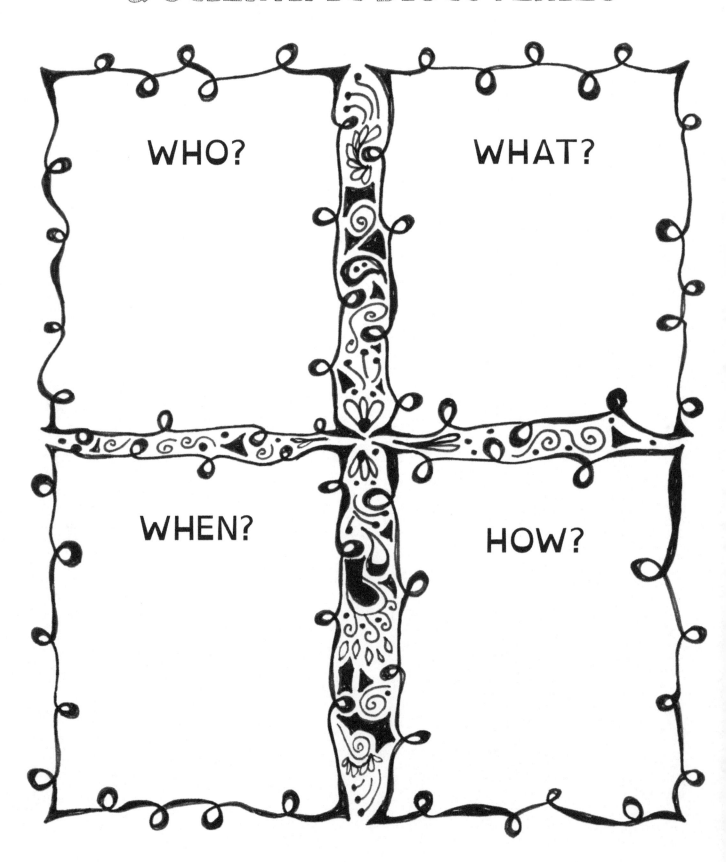

WHO?

WHAT?

WHEN?

HOW?

Drawings, Photos & Images:

A STUDY OF SCIENTISTS & SCIENTIFIC DISCOVERIES

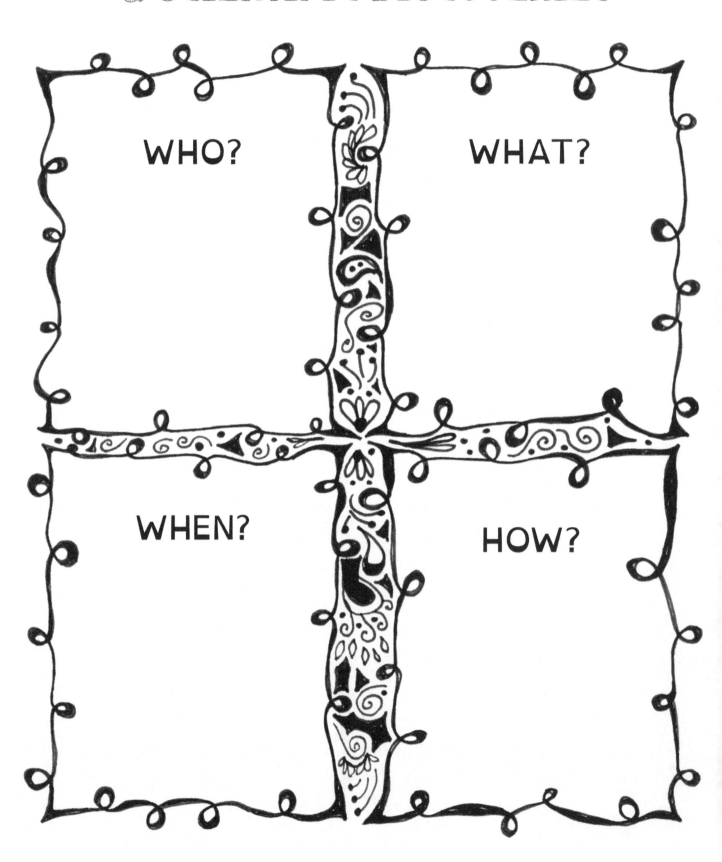

WHO?

WHAT?

WHEN?

HOW?

Drawings, Photos & Images:

A STUDY OF SCIENTISTS & SCIENTIFIC DISCOVERIES

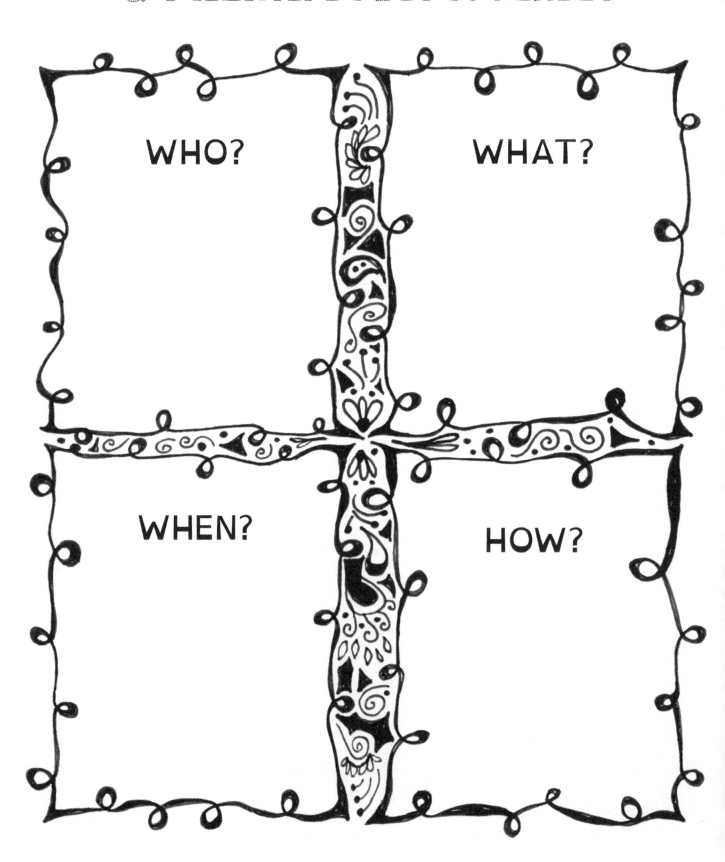

WHO?

WHAT?

WHEN?

HOW?

Drawings, Photos & Images:

A STUDY OF SCIENTISTS & SCIENTIFIC DISCOVERIES

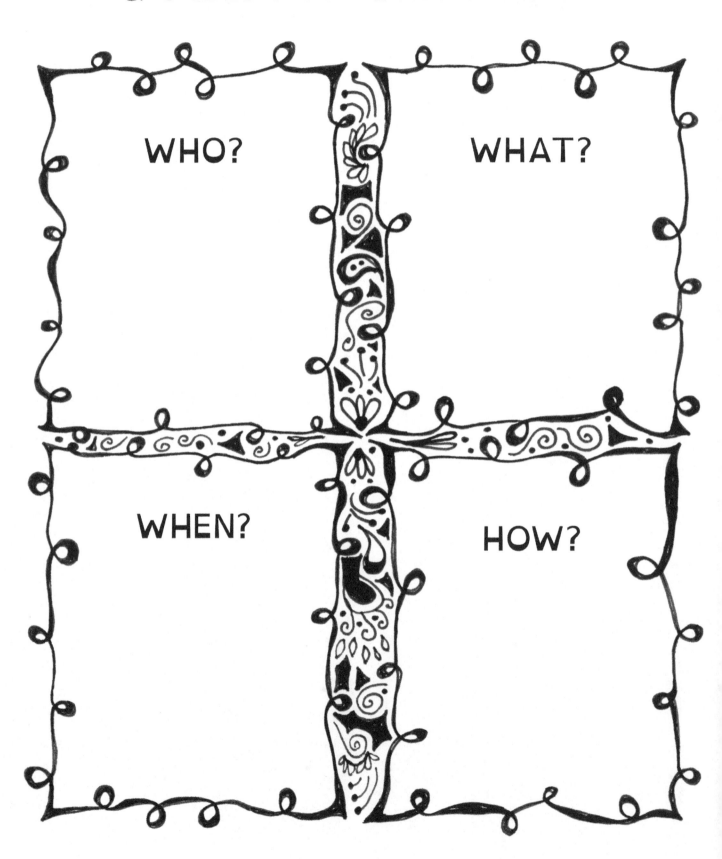

WHO?

WHAT?

WHEN?

HOW?

Drawings, Photos & Images:

SCIENCE VOCABULARY

Date:_____

Look-up and Define 5 Words:

1. _____

_____

_____

2. _____

_____

_____

3. _____

_____

_____

4. _____

_____

_____

5. _____

_____

_____

Choose 10 Spelling Words
Related to Science

1_____

2_____

3_____

4_____

5_____

6_____

7_____

8_____

9_____

10_____

SCIENCE VOCABULARY

Date:_____

Look-up and Define 5 Words:

1. _____

_____

_____

2. _____

_____

_____

3. _____

_____

_____

4. _____

_____

_____

5. _____

_____

_____

Choose 10 Spelling Words

Related to Science

1_____

2_____

3_____

4_____

5_____

6_____

7_____

8_____

9_____

10_____

SCIENCE VOCABULARY

Date:_____

Look-up and Define 5 Words:

1.

2.

3.

4.

5.

Choose 10 Spelling Words
Related to Science

1_____

2_____

3_____

4_____

5_____

6_____

7_____

8_____

9_____

10_____

SCIENCE VOCABULARY

Date:_____

Look-up and Define 5 Words:

1. _____

_____

_____

2. _____

_____

_____

3. _____

_____

4. _____

_____

_____

5. _____

_____

_____

Choose 10 Spelling Words
Related to Science

1_____

2_____

3_____

4_____

5_____

6_____

7_____

8_____

9_____

10_____

76

SCIENCE VOCABULARY

Date:_____

Look-up and Define 5 Words:

1.

2.

3.

4.

5.

Choose 10 Spelling Words Related to Science

1_____

2_____

3_____

4_____

5_____

6_____

7_____

8_____

9_____

10_____

SCIENCE FILMS & TUTORIALS

List all the Films, Tutorials & Documentaries

That you are using for your research and training:

DATE: _____ TITLE: _____

TOPIC: _____

_____

DATE: _____ TITLE: _____

TOPIC: _____

_____

DATE: _____ TITLE: _____

TOPIC: _____

_____

DATE: _____ TITLE: _____

TOPIC: _____

_____

DATE: _____ TITLE: _____

TOPIC: _____

_____

DATE: _____ TITLE: _____

TOPIC: _____

_____

DATE:

Title:

_____

Topic:_____

Illustrations:

Notes:_____
_____
_____
_____
_____

Comments:

Science Films & Tutorials

List all the Films, Tutorials & Documentaries

That you are using for your research and training:

DATE: _____TITLE: _____

TOPIC:_____

_____

DATE: _____TITLE: _____

TOPIC:_____

_____

DATE: _____TITLE: _____

TOPIC:_____

_____

DATE: _____TITLE: _____

TOPIC:_____

_____

DATE: _____TITLE: _____

TOPIC:_____

_____

DATE: _____TITLE: _____

TOPIC:_____

_____

DATE:

Title:

_____

Topic:_____

Illustrations:

Notes:_____

_____

_____

_____

_____

Comments:

Science Films & Tutorials

List all the Films, Tutorials & Documentaries

That you are using for your research and training:

DATE: _____TITLE: _____

TOPIC:_____

_____

DATE: _____TITLE: _____

TOPIC:_____

_____

DATE: _____TITLE: _____

TOPIC:_____

_____

DATE: _____TITLE: _____

TOPIC:_____

_____

DATE: _____TITLE: _____

TOPIC:_____

_____

DATE: _____TITLE: _____

TOPIC:_____

_____

DATE:

Title:

_____

Topic:_____

Illustrations:

Notes:_____

_____

_____

_____

_____

Comments:

Science Films & Tutorials

List all the Films, Tutorials & Documentaries

That you are using for your research and training:

DATE: _____TITLE: _____

TOPIC:_____

_____

DATE: _____TITLE: _____

TOPIC:_____

_____

DATE: _____TITLE: _____

TOPIC:_____

_____

DATE: _____TITLE: _____

TOPIC:_____

_____

DATE: _____TITLE: _____

TOPIC:_____

_____

DATE: _____TITLE: _____

TOPIC:_____

_____

DATE:

Title:

_____

Topic:_____

Illustrations:

Notes:_____

_____

_____

_____

_____

Comments:

Science Films & Tutorials

List all the Films, Tutorials & Documentaries

That you are using for your research and training:

DATE: _____TITLE: _____

TOPIC:_____

_____

DATE: _____TITLE: _____

TOPIC:_____

_____

DATE: _____TITLE: _____

TOPIC:_____

_____

DATE: _____TITLE: _____

TOPIC:_____

_____

DATE: _____TITLE: _____

TOPIC:_____

_____

DATE: _____TITLE: _____

TOPIC:_____

_____

DATE:

Title:

_____

Topic:_____

Illustrations:

Notes:_____

_____

_____

_____

_____

Comments:

MY SCIENCE PROJECTS

Experiment. Make Something. Build Something. Invent Something.

Cook Something. Investigate Something. Design Something.

Plan Something. Do a Project.

Draw pictures of your projects or tape photos in this section.

List of Projects:

1 _____

_____

_____

2 _____

_____

_____

3 _____

_____

_____

4 _____

_____

_____

List of Projects:

6 _____

_____

_____

7 _____

_____

_____

8 _____

_____

_____

9 _____

_____

_____

10 _____

_____

_____

11 _____

_____

_____

SCIENCE PROJECT

Name of Project:

_____

Date Started: _____ Date Complete:_____

Materials:_____

_____

_____

Results:_____

_____

_____

Images:

Drawings, Photos & Images:

SCIENCE PROJECT

#2

Name of Project:

_____

Date Started: _____ Date Complete:_____

Materials:_____

_____

_____

Results:_____

_____

_____

Images:

Drawings, Photos & Images:

SCIENCE PROJECT

Name of Project:

_____

Date Started: _____ Date Complete:_____

Materials:_____

_____

_____

Results:_____

_____

_____

Images:

Drawings, Photos & Images:

SCIENCE PROJECT

#4

Name of Project:

_____

Date Started: _____ Date Complete: _____

Materials: _____

_____

_____

Results: _____

_____

_____

Images:

Drawings, Photos & Images:

SCIENCE PROJECT

Name of Project:

_____

Date Started: _____Date Complete:_____

Materials:_____

_____

_____

Results:_____

_____

_____

Images:

Drawings, Photos & Images:

SCIENCE PROJECT

Name of Project:

_____

Date Started: _____Date Complete:_____

Materials:_____

_____

_____

Results:_____

_____

_____

Images:

Drawings, Photos & Images:

SCIENCE PROJECT

Name of Project:

_____

Date Started: _____ Date Complete:_____

Materials:_____

_____

_____

Results:_____

_____

_____

Images:

Drawings, Photos & Images:

SCIENCE PROJECT

Name of Project:

_____

Date Started: _____ Date Complete:_____

Materials:_____

_____

_____

Results:_____

_____

_____

Images:

Drawings, Photos & Images:

SCIENCE PROJECT

Name of Project:

Date Started: _____ Date Complete:_____

Materials:_____

_____

_____

Results:_____

_____

_____

Images:

Drawings, Photos & Images:

SCIENCE PROJECT

Name of Project:

_____

Date Started: _____ Date Complete:_____

Materials:_____

_____

_____

Results:_____

_____

_____

Images:

Drawings, Photos & Images:

SCIENCE PROJECT

Name of Project:

_____

Date Started: _____ Date Complete:_____

Materials:_____

_____

_____

Results:_____

_____

_____

Images:

Drawings, Photos & Images:

SCIENCE PROJECT

#12

Name of Project:

_____

Date Started: _____ Date Complete: _____

Materials: _____

_____

_____

Results: _____

_____

_____

Images:

Drawings, Photos & Images:

RECORD KEEPING, CHARTS & GRAPHS

RECORD KEEPING, CHARTS & GRAPHS

RECORD KEEPING, CHARTS & GRAPHS

RECORD KEEPING, CHARTS & GRAPHS

RECORD KEEPING, CHARTS & GRAPHS

RECORD KEEPING, CHARTS & GRAPHS

RECORD KEEPING, CHARTS & GRAPHS

RECORD KEEPING, CHARTS & GRAPHS

RECORD KEEPING, CHARTS & GRAPHS

RECORD KEEPING, CHARTS & GRAPHS

RESEARCH, ESSAYS, AND CREATIVE WRITING

Use these pages for research

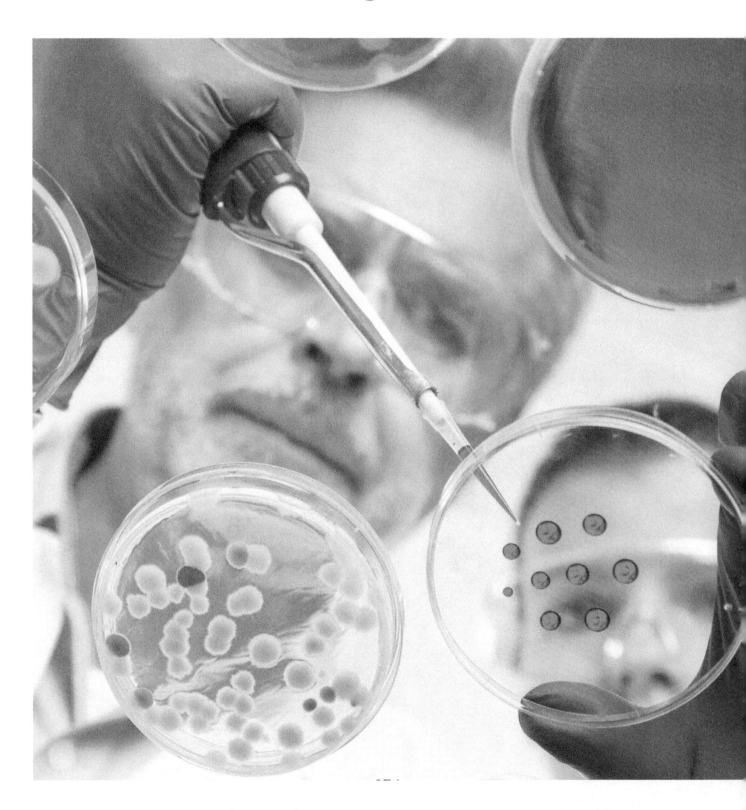

TITLE:_____ Date:_____

RESEARCH & WRITING

TITLE:_____ Date:_____

RESEARCH & WRITING

TITLE:_____ Date:_____

RESEARCH & WRITING

TITLE:_____ Date:_____

RESEARCH & WRITING

TITLE:_____ Date:_____

RESEARCH & WRITING

TITLE:_____ Date:_____

RESEARCH & WRITING

TITLE:_____ Date:_____

RESEARCH & WRITING

TITLE:_____ Date:_____

RESEARCH & WRITING

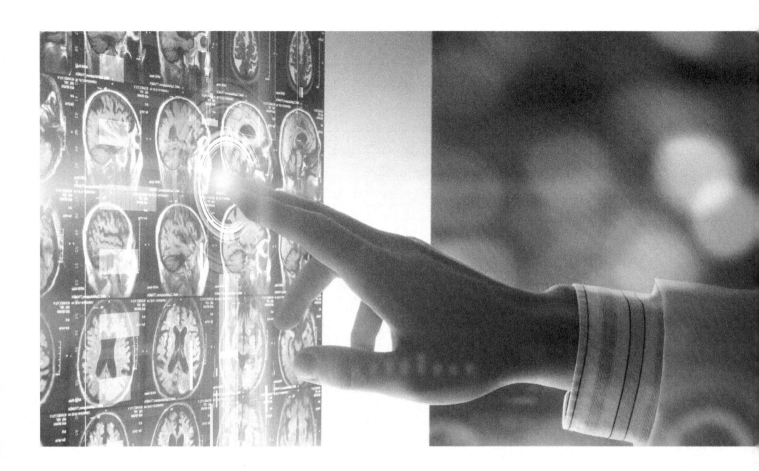

TITLE:_____ Date:_____

RESEARCH & WRITING

TITLE:_____ Date:_____

143

RESEARCH & WRITING

TITLE:_____ Date:_____

RESEARCH & WRITING

TITLE:_____ Date:_____

RESEARCH & WRITING

TITLE:_____ Date:_____

RESEARCH & WRITING

TITLE:_____ Date:_____

RESEARCH & WRITING

TITLE:_____ Date:_____

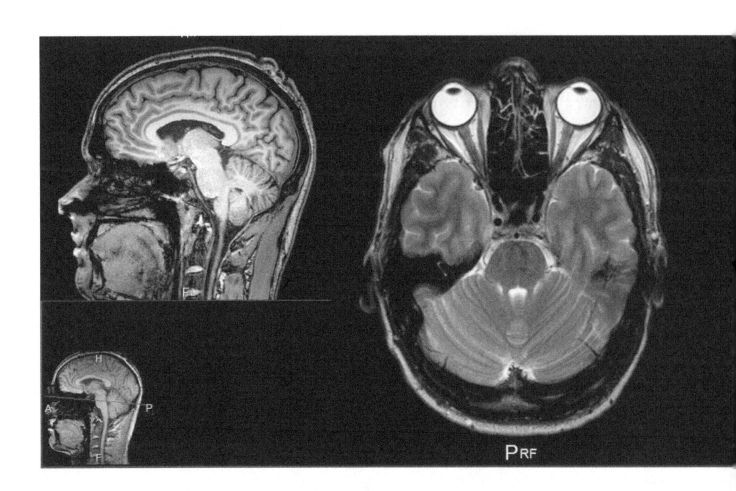

TITLE:_____ Date:_____

RESEARCH & WRITING

TITLE:_____ Date:_____

RESEARCH & WRITING

TITLE:_____ Date:_____

RESEARCH & WRITING

TITLE:_____ Date:_____

RESEARCH & WRITING

TITLE:_____ Date:_____

RESEARCH & WRITING

TITLE:_____ Date:_____

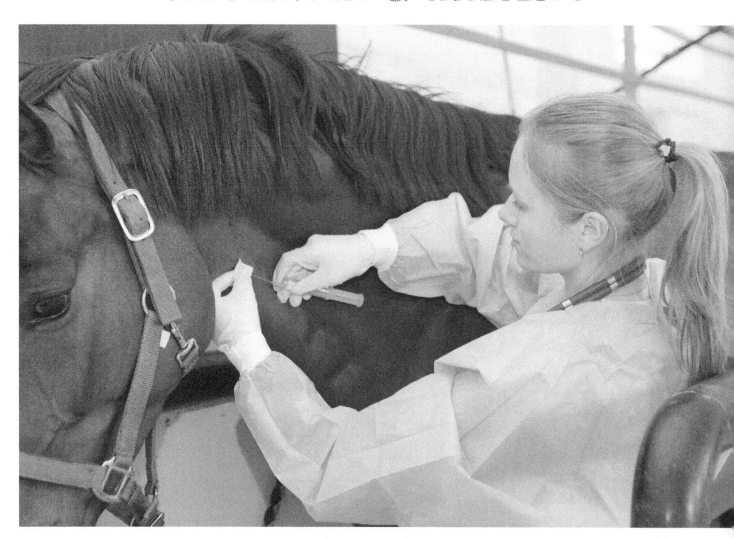

TITLE:_____ Date:_____

RESEARCH & WRITING

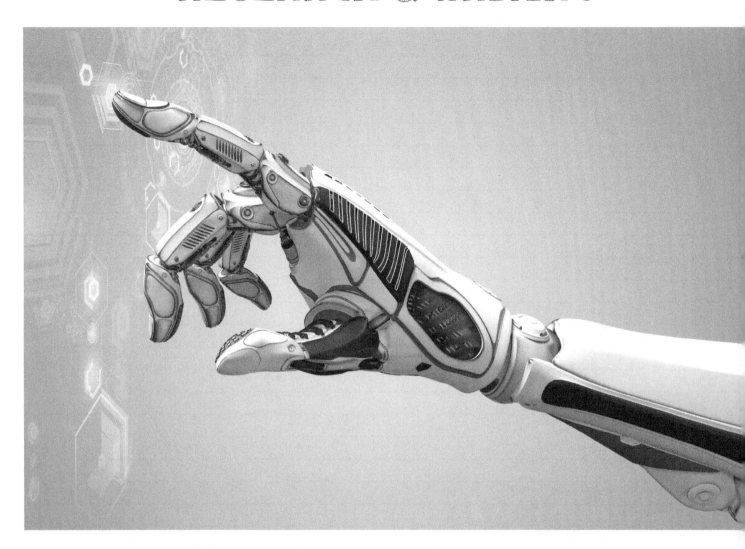

TITLE:_____ Date:_____

RESEARCH & WRITING

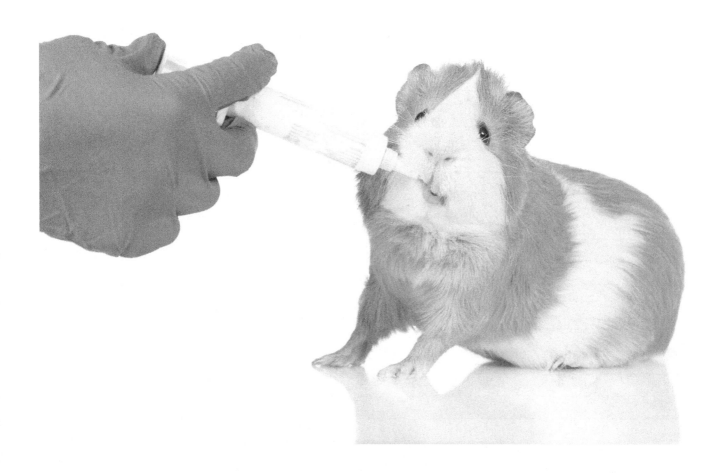

TITLE:_____ Date:_____

RESEARCH & WRITING

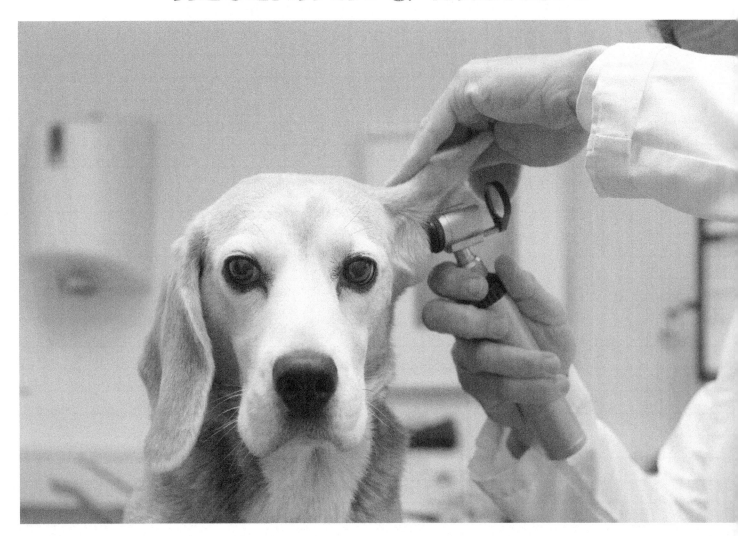

TITLE:_____ Date:_____

RESEARCH & WRITING

TITLE:_____ Date:_____

TITLE:_____ Date:_____

PERIODIC TABLE OF ELEMENTS

1
H
1.0079
Hydrogen

3	4
Li	**Be**
1.941	9.0122
Lithium	Beryllium

11	12
Na	**Mg**
22.990	24.305
Sodium	Magnesium

19	20	21	22	23	24	25	26	27
K	**Ca**	**Sc**	**Ti**	**V**	**Cr**	**Mn**	**Fe**	**Co**
39.098	40.078	44.956	47.867	50.942	51.996	54.938	55.845	58.933
Potassium	Calcium	Scandium	Titanium	Vanadium	Chromium	Manganese	Iron	Cobalt

37	38	39	40	41	42	43	44	45
Rb	**Sr**	**Y**	**Zr**	**Nb**	**Mo**	**Tc**	**Ru**	**Rh**
85.468	87.62	88.906	91.224	92.906	95.94	98	101.07	102.91
Rubidium	Strontium	Yttrium	Zirconium	Niobium	Molybdenum	Technetium	Ruthenium	Rhodium

55	56	57 - 71	72	73	74	75	76	77
Cs	**Ba**	**La - Lu**	**Hf**	**Ta**	**W**	**Re**	**Os**	**Ir**
132.91	137.33		178.49	180.95	183.84	186.21	190.23	192.22
Cesium	Barium		Hafnium	Tantalum	Tungsten	Rhenium	Osmium	Iridium

87	88	89 - 103	104	105	106	107	108	109
Fr	**Ra**	**Ac - Lr**	**Rf**	**Db**	**Sg**	**Bh**	**Hs**	**Mt**
223	226		261	262	266	264	269	268
Francium	Radium		Rutherfordium	Dubnium	Seaborgium	Bohrium	Hassium	Meitnerium

Lanthanide series

57	58	59	60	61	62	63
La	**Ce**	**Pr**	**Nd**	**Pm**	**Sm**	**Eu**
138.91	140.12	140.91	144.24	145	150.36	151.96
Lanthanide	Cerium	Praseodymium	Neodymium	Promethium	Samarium	Europium

Actinide series

89	90	91	92	93	94	95
Ac	**Th**	**Pa**	**U**	**Np**	**Pu**	**Am**
227	232.04	231.04	238.03	237	244	243
Actinide	Thorium	Protactinium	Uranium	Neptunium	Plutonium	Americium

											2 **He** 4.0026 Helium

5 **B** 10.811 Boron	6 **C** 12.011 Carbon	7 **N** 14.007 Nitrogen	8 **O** 15.999 Oxygen	9 **F** 18.998 Fluorine	10 **Ne** 20.180 Neon
13 **Al** 26.982 Aluminium	14 **Si** 28.086 Silicon	15 **P** 30.974 Phosphorus	16 **S** 32.065 Sulfur	17 **Cl** 35.453 Chlorine	18 **Ar** 39.948 Argon

28 **Ni** 58.693 Nickel	29 **Cu** 63.546 Cooper	30 **Zn** 65.39 Zinc	31 **Ga** 69.723 Gallium	32 **Ge** 1.0079 Germanium	33 **As** 74.992 Arsenic	34 **Se** 78.96 Selenium	35 **Br** 79.904 Bromine	36 **Kr** 83.80 Krypton
46 **Pd** 106.42 Palladium	47 **Ag** 107.87 Silver	48 **Cd** 112.41 Cadmium	49 **In** 114.82 Indium	50 **Sn** 118.71 Tin	51 **Sb** 121.76 Antimony	52 **Te** 127.60 Tellurium	53 **I** 126.90 Iodine	54 **Xe** 131.29 Xenon
78 **Pt** 195.08 Platinum	79 **Au** 196.97 Gold	80 **Hg** 200.59 Mercury	81 **Tl** 204.38 Thallium	82 **Pb** 207.2 Lead	83 **Bi** 208.98 Bismuth	84 **Po** 209 Polonium	85 **At** 210 Astatine	86 **Rn** 222 Radon
110 **Uun** 271 Ununnilium	111 **Uuu** 272 Unununium	112 **Uub** 1.0079 Ununbium	113 **Uut** Ununtrium	114 **Uuq** 289 Ununquadium	115 **Uup** Ununpentium	116 **Uuh** Ununhexium	117 **Uus** Ununseptium	118 **Uuo** Ununoctium

64 **Gd** 157.25 Gadolinum	65 **Tb** 158.93 Terbium	66 **Dy** 162.5 Dysprosium	67 **Ho** 164.93 Holmium	68 **Er** 1.0079 Erbium	69 **Tm** 168.93 Thulium	70 **Yb** 173.04 Yttersium	71 **Lu** 1.0079 Lutetium
96 **Cm** 247 Curium	97 **Bk** 247 Berkelium	98 **Cf** 251 Californium	99 **Es** 252 Einsteinium	100 **Fm** 257 Fermium	101 **Md** 258 Mendelevium	102 **No** 259 Nobelium	103 **Lr** 1.0079 Lawrencium

MAKE YOUR OWN
PERIODIC TABLE OF ELEMENTS

PERIODIC TABLE OF ELEMENTS

1 **H** 1.0079 Hydrogen										

3 **Li** 1.941 Lithium	4 **Be** 9.0122 Beryllium

11 **Na** 22.990 Sodium	12 **Mg** 24.305 Magnesium

19 **K** 39.098 Potassium	20 **Ca** 40.078 Calcium	21 **Sc** 44.956 Scandium	22 **Ti** 47.867 Titanium	23 **V** 50.942 Vanadium	24 **Cr** 51.996 Chromium	25 **Mn** 54.938 Manganese	26 **Fe** 55.845 Iron	27 **Co** 58.933 Cobalt
37 **Rb** 85.468 Rubidium	38 **Sr** 87.62 Strontium	39 **Y** 88.906 Yttrium	40 **Zr** 91.224 Zirconium	41 **Nb** 92.906 Niobium	42 **Mo** 95.94 Molybdenum	43 **Tc** 98 Technetium	44 **Ru** 101.07 Ruthenium	45 **Rh** 102.91 Rhodium
55 **Cs** 132.91 Cesium	56 **Ba** 137.33 Barium	57 - 71 **La-Lu**	72 **Hf** 178.49 Hafnium	73 **Ta** 180.95 Tantalum	74 **W** 183.84 Tungsten	75 **Re** 186.21 Rhenium	76 **Os** 190.23 Osmium	77 **Ir** 192.22 Iridium
87 **Fr** 223 Francium	88 **Ra** 226 Radium	89 - 103 **Ac-Lr**	104 **Rf** 261 Rutherfordium	105 **Db** 262 Dubnium	106 **Sg** 266 Seaborgium	107 **Bh** 264 Bohrium	108 **Hs** 269 Hassium	109 **Mt** 268 Meitnerium

Lanthanide series	57 **La** 138.91 Lanthanide	58 **Ce** 140.12 Cerium	59 **Pr** 140.91 Praseodymium	60 **Nd** 144.24 Neodymium	61 **Pm** 145 Promethium	62 **Sm** 150.36 Samarium	63 **Eu** 151.96 Europium
Actinide series	89 **Ac** 227 Actinide	90 **Th** 232.04 Thorium	91 **Pa** 231.04 Protactinium	92 **U** 238.03 Uranium	93 **Np** 237 Neptunium	94 **Pu** 244 Plutonium	95 **Am** 243 Americium

THE ELEMENTS

						2 **He** 4.0026 Helium

5 **B** 10.811 Boron	6 **C** 12.011 Carbon	7 **N** 14.007 Nitrogen	8 **O** 15.999 Oxygen	9 **F** 18.998 Fluorine	10 **Ne** 20.180 Neon

13 **Al** 26.982 Aluminium	14 **Si** 28.086 Silicon	15 **P** 30.974 Phosphorus	16 **S** 32.065 Sulfur	17 **Cl** 35.453 Chlorine	18 **Ar** 39.948 Argon

28 **Ni** 58.693 Nickel	29 **Cu** 63.546 Cooper	30 **Zn** 65.39 Zinc	31 **Ga** 69.723 Gallium	32 **Ge** 1.0079 Germanium	33 **As** 74.992 Arsenic	34 **Se** 78.96 Selenium	35 **Br** 79.904 Bromine	36 **Kr** 83.80 Krypton

46 **Pd** 106.42 Palladium	47 **Ag** 107.87 Silver	48 **Cd** 112.41 Cadmium	49 **In** 114.82 Indium	50 **Sn** 118.71 Tin	51 **Sb** 121.76 Antimony	52 **Te** 127.60 Tellurium	53 **I** 126.90 Iodine	54 **Xe** 131.29 Xenon

78 **Pt** 195.08 Platinum	79 **Au** 196.97 Gold	80 **Hg** 200.59 Mercury	81 **Tl** 204.38 Thallium	82 **Pb** 207.2 Lead	83 **Bi** 208.98 Bismuth	84 **Po** 209 Polonium	85 **At** 210 Astatine	86 **Rn** 222 Radon

110 **Uun** 271 Ununnilium	111 **Uuu** 272 Unununium	112 **Uub** 1.0079 Ununbium	113 **Uut** Ununtrium	114 **Uuq** 289 Ununquadium	115 **Uup** Ununpentium	116 **Uuh** Ununhexium	117 **Uus** Ununseptium	118 **Uuo** Ununoctium

64 **Gd** 157.25 Gadolinum	65 **Tb** 158.93 Terbium	66 **Dy** 162.5 Dysprosium	67 **Ho** 164.93 Holmium	68 **Er** 1.0079 Erbium	69 **Tm** 168.93 Thulium	70 **Yb** 173.04 Yttersium	71 **Lu** 1.0079 Lutetium

96 **Cm** 247 Curium	97 **Bk** 247 Berkelium	98 **Cf** 251 Californium	99 **Es** 252 Einsteinium	100 **Fm** 257 Fermium	101 **Md** 258 Mendelevium	102 **No** 259 Nobelium	103 **Lr** 1.0079 Lawrencium

MAKE YOUR OWN
PERIODIC TABLE OF ELEMENTS

PERIODIC TABLE OF ELEMENTS

1 **H** 1.0079 Hydrogen								
3 **Li** 1.941 Lithium	**4** **Be** 9.0122 Beryllium							
11 **Na** 22.990 Sodium	**12** **Mg** 24.305 Magnesium							
19 **K** 39.098 Potassium	**20** **Ca** 40.078 Calcium	**21** **Sc** 44.956 Scandium	**22** **Ti** 47.867 Titanium	**23** **V** 50.942 Vanadium	**24** **Cr** 51.996 Chromium	**25** **Mn** 54.938 Manganese	**26** **Fe** 55.845 Iron	**27** **Co** 58.933 Cobalt
37 **Rb** 85.468 Rubidium	**38** **Sr** 87.62 Strontium	**39** **Y** 88.906 Yttrium	**40** **Zr** 91.224 Zirconium	**41** **Nb** 92.906 Niobium	**42** **Mo** 95.94 Molybdenum	**43** **Tc** 98 Technetium	**44** **Ru** 101.07 Ruthenium	**45** **Rh** 102.91 Rhodium
55 **Cs** 132.91 Cesium	**56** **Ba** 137.33 Barium	**57 - 71** **La - Lu**	**72** **Hf** 178.49 Hafnium	**73** **Ta** 180.95 Tantalum	**74** **W** 183.84 Tungsten	**75** **Re** 186.21 Rhenium	**76** **Os** 190.23 Osmium	**77** **Ir** 192.22 Iridium
87 **Fr** 223 Francium	**88** **Ra** 226 Radium	**89 - 103** **Ac - Lr**	**104** **Rf** 261 Rutherfordium	**105** **Db** 262 Dubnium	**106** **Sg** 266 Seaborgium	**107** **Bh** 264 Bohrium	**108** **Hs** 269 Hassium	**109** **Mt** 268 Meitnerium

Lanthanide series	**57** **La** 138.91 Lanthanide	**58** **Ce** 140.12 Cerium	**59** **Pr** 140.91 Praseodymium	**60** **Nd** 144.24 Neodymium	**61** **Pm** 145 Promethium	**62** **Sm** 150.36 Samarium	**63** **Eu** 151.96 Europium
Actinide series	**89** **Ac** 227 Actinide	**90** **Th** 232.04 Thorium	**91** **Pa** 231.04 Protactinium	**92** **U** 238.03 Uranium	**93** **Np** 237 Neptunium	**94** **Pu** 244 Plutonium	**95** **Am** 243 Americium

THE ELEMENTS

								2 **He** 4.0026 Helium

5 **B** 10.811 Boron	6 **C** 12.011 Carbon	7 **N** 14.007 Nitrogen	8 **O** 15.999 Oxygen	9 **F** 18.998 Fluorine	10 **Ne** 20.180 Neon

13 **Al** 26.982 Aluminium	14 **Si** 28.086 Silicon	15 **P** 30.974 Phosphorus	16 **S** 32.065 Sulfur	17 **Cl** 35.453 Chlorine	18 **Ar** 39.948 Argon

28 **Ni** 58.693 Nickel	29 **Cu** 63.546 Cooper	30 **Zn** 65.39 Zinc	31 **Ga** 69.723 Gallium	32 **Ge** 1.0079 Germanium	33 **As** 74.992 Arsenic	34 **Se** 78.96 Selenium	35 **Br** 79.904 Bromine	36 **Kr** 83.80 Krypton

46 **Pd** 106.42 Palladium	47 **Ag** 107.87 Silver	48 **Cd** 112.41 Cadmium	49 **In** 114.82 Indium	50 **Sn** 118.71 Tin	51 **Sb** 121.76 Antimony	52 **Te** 127.60 Tellurium	53 **I** 126.90 Iodine	54 **Xe** 131.29 Xenon

78 **Pt** 195.08 Platinum	79 **Au** 196.97 Gold	80 **Hg** 200.59 Mercury	81 **Tl** 204.38 Thallium	82 **Pb** 207.2 Lead	83 **Bi** 208.98 Bismuth	84 **Po** 209 Polonium	85 **At** 210 Astatine	86 **Rn** 222 Radon

110 **Uun** 271 Ununnilium	111 **Uuu** 272 Unununium	112 **Uub** 1.0079 Ununbium	113 **Uut** Ununtrium	114 **Uuq** 289 Ununquadium	115 **Uup** Ununpentium	116 **Uuh** Ununhexium	117 **Uus** Ununseptium	118 **Uuo** Ununoctium

64 **Gd** 157.25 Gadolinum	65 **Tb** 158.93 Terbium	66 **Dy** 162.5 Dysprosium	67 **Ho** 164.93 Holmium	68 **Er** 1.0079 Erbium	69 **Tm** 168.93 Thulium	70 **Yb** 173.04 Yttersium	71 **Lu** 1.0079 Lutetium

96 **Cm** 247 Curium	97 **Bk** 247 Berkelium	98 **Cf** 251 Californium	99 **Es** 252 Einsteinium	100 **Fm** 257 Fermium	101 **Md** 258 Mendelevium	102 **No** 259 Nobelium	103 **Lr** 1.0079 Lawrencium

MAKE YOUR OWN
PERIODIC TABLE OF ELEMENTS

PERIODIC TABLE OF ELEMENTS

1 **H** 1.0079 Hydrogen									
3 **Li** 1.941 Lithium	4 **Be** 9.0122 Beryllium								
11 **Na** 22.990 Sodium	12 **Mg** 24.305 Magnesium								
19 **K** 39.098 Potassium	20 **Ca** 40.078 Calcium	21 **Sc** 44.956 Scandium	22 **Ti** 47.867 Titanium	23 **V** 50.942 Vanadium	24 **Cr** 51.996 Chromium	25 **Mn** 54.938 Manganese	26 **Fe** 55.845 Iron	27 **Co** 58.933 Cobalt
37 **Rb** 85.468 Rubidium	38 **Sr** 87.62 Strontium	39 **Y** 88.906 Yttrium	40 **Zr** 91.224 Zirconium	41 **Nb** 92.906 Niobium	42 **Mo** 95.94 Molybdenum	43 **Tc** 98 Technetium	44 **Ru** 101.07 Ruthenium	45 **Rh** 102.91 Rhodium
55 **Cs** 132.91 Cesium	56 **Ba** 137.33 Barium	57 - 71 **La - Lu**	72 **Hf** 178.49 Hafnium	73 **Ta** 180.95 Tantalum	74 **W** 183.84 Tungsten	75 **Re** 186.21 Rhenium	76 **Os** 190.23 Osmium	77 **Ir** 192.22 Iridium
87 **Fr** 223 Francium	88 **Ra** 226 Radium	89 - 103 **Ac - Lr**	104 **Rf** 261 Rutherfordium	105 **Db** 262 Dubnium	106 **Sg** 266 Seaborgium	107 **Bh** 264 Bohrium	108 **Hs** 269 Hassium	109 **Mt** 268 Meitnerium

Lanthanide series

57 **La** 138.91 Lanthanide	58 **Ce** 140.12 Cerium	59 **Pr** 140.91 Praseodymium	60 **Nd** 144.24 Neodymium	61 **Pm** 145 Promethium	62 **Sm** 150.36 Samarium	63 **Eu** 151.96 Europium

Actinide series

89 **Ac** 227 Actinide	90 **Th** 232.04 Thorium	91 **Pa** 231.04 Protactinium	92 **U** 238.03 Uranium	93 **Np** 237 Neptunium	94 **Pu** 244 Plutonium	95 **Am** 243 Americium

THE ELEMENTS

2
He
4.0026
Helium

5	6	7	8	9	10
B	**C**	**N**	**O**	**F**	**Ne**
10.811	12.011	14.007	15.999	18.998	20.180
Boron	Carbon	Nitrogen	Oxygen	Fluorine	Neon

13	14	15	16	17	18
Al	**Si**	**P**	**S**	**Cl**	**Ar**
26.982	28.086	30.974	32.065	35.453	39.948
Aluminium	Silicon	Phosphorus	Sulfur	Chlorine	Argon

28	29	30	31	32	33	34	35	36
Ni	**Cu**	**Zn**	**Ga**	**Ge**	**As**	**Se**	**Br**	**Kr**
58.693	63.546	65.39	69.723	1.0079	74.992	78.96	79.904	83.80
Nickel	Cooper	Zinc	Gallium	Germanium	Arsenic	Selenium	Bromine	Krypton

46	47	48	49	50	51	52	53	54
Pd	**Ag**	**Cd**	**In**	**Sn**	**Sb**	**Te**	**I**	**Xe**
106.42	107.87	112.41	114.82	118.71	121.76	127.60	126.90	131.29
Palladium	Silver	Cadmium	Indium	Tin	Antimony	Tellurium	Iodine	Xenon

78	79	80	81	82	83	84	85	86
Pt	**Au**	**Hg**	**Tl**	**Pb**	**Bi**	**Po**	**At**	**Rn**
195.08	196.97	200.59	204.38	207.2	208.98	209	210	222
Platinum	Gold	Mercury	Thallium	Lead	Bismuth	Polonium	Astatine	Radon

110	111	112	113	114	115	116	117	118
Uun	**Uuu**	**Uub**	**Uut**	**Uuq**	**Uup**	**Uuh**	**Uus**	**Uuo**
271	272	1.0079		289				
Ununnilium	Unununium	Ununbium	Ununtrium	Ununquadium	Ununpentium	Ununhexium	Ununseptium	Ununoctium

64	65	66	67	68	69	70	71
Gd	**Tb**	**Dy**	**Ho**	**Er**	**Tm**	**Yb**	**Lu**
157.25	158.93	162.5	164.93	1.0079	168.93	173.04	1.0079
Gadolinum	Terbium	Dysprosium	Holmium	Erbium	Thulium	Yttersium	Lutetium

96	97	98	99	100	101	102	103
Cm	**Bk**	**Cf**	**Es**	**Fm**	**Md**	**No**	**Lr**
247	247	251	252	257	258	259	1.0079
Curium	Berkelium	Californium	Einsteinium	Fermium	Mendelevium	Nobelium	Lawrencium

MAKE YOUR OWN
PERIODIC TABLE OF ELEMENTS

Draw the Covers of All Your Extra
SCIENCE BOOKS

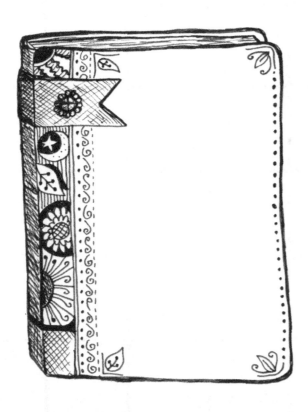

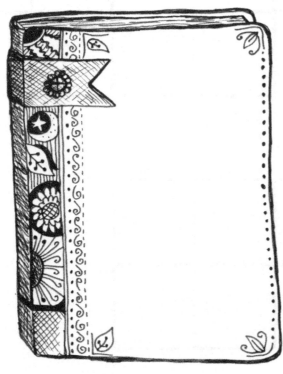
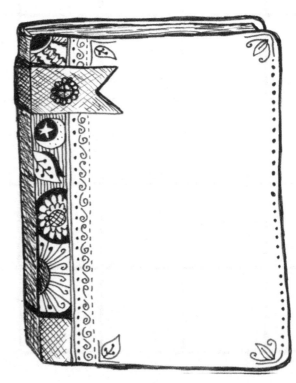

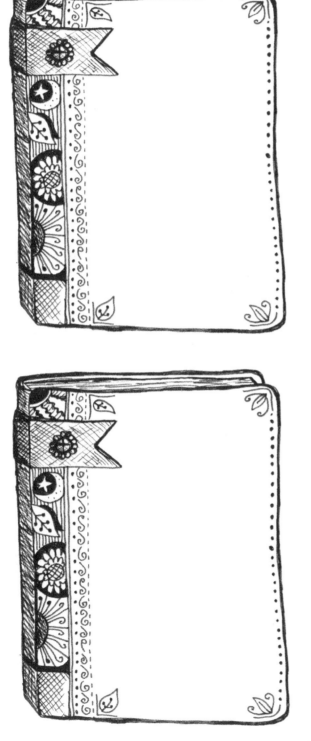

EXTRA BOOKS

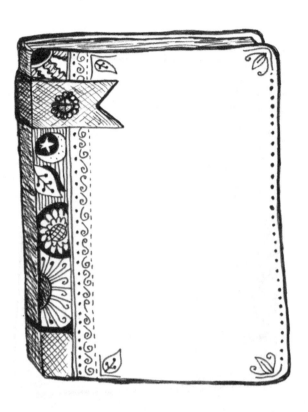

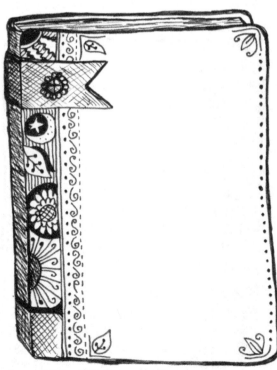

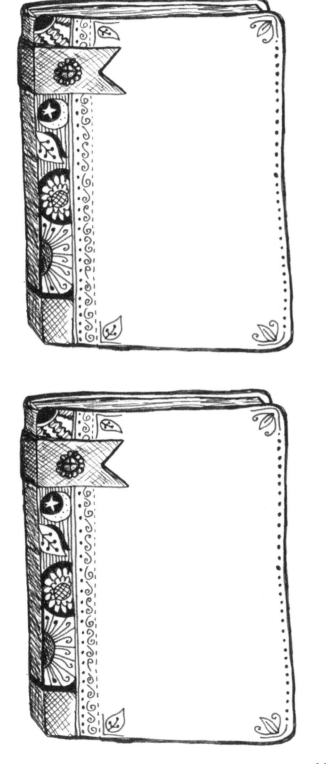

EXTRA BOOKS

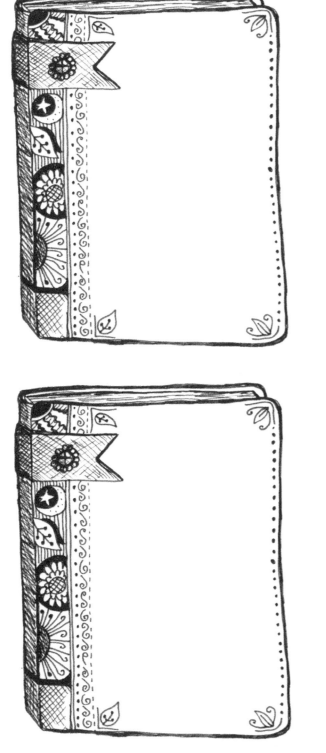

Made in United States
Orlando, FL
03 August 2023

35716085R00111